SPOTLIGHT ON NATURE
CHEETAH

LAURA PURDIE SALAS

CREATIVE EDUCATION · CREATIVE PAPERBACKS

Published by Creative Education and Creative Paperbacks
P.O. Box 227, Mankato, Minnesota 56002
Creative Education and Creative Paperbacks are imprints
of The Creative Company
www.thecreativecompany.us

Design by Rhea Magaro; production by Beeline Media & Design, Inc.
Art direction by Tom Morgan
Printed in the United States of America

Photographs by Alamy (Rob Schultz, Ann and Steve Toon, Lennjo), Getty Images (Matt Polski, Anup Shah, wilpunt), Shutterstock (Planetphoto.ch, Le Do, Julian W, Sunny_nsk, Victor Lapaev, slowmotiongli, Lost Al, stuiesk, GUDKOV ANDREY, Stu Porter, Hedrus, Mateo Juric, Maggy Meyer, Kevin Wells Photography, nwdph, Wirestock Creators)

Library of Congress Cataloging-in-Publication Data
Names: Salas, Laura Purdie, author.
Title: Cheetah / Laura Purdie Salas.
Description: Mankato, Minnesota : Creative Education and Creative Paperbacks, [2025] | Series: Spotlight on nature | Includes bibliographical references and index. | Audience: Ages 10-13 | Audience: Grades 4-6 | Summary: "Dash into the cheetah's world with our new book for mid-level readers. An engaging narrative of a cheetah family, captivating images, infographics, milestones, and resources make it perfect for young nature enthusiasts"-- Provided by publisher.
Identifiers: LCCN 2023059416 (print) | LCCN 2023059417 (ebook) | ISBN 9798889891901 (library binding) | ISBN 9781682775639 (paperback) | ISBN 9798889891987 (ebook)
Subjects: LCSH: Cheetah--Juvenile literature. | Cheetah--Life cycles--Juvenile literature. | Cheetah--Zimbabwe--Hwange National Park--Juvenile literature.
Classification: LCC QL737.C23 S228 2025 (print) | LCC QL737.C23 (ebook) |
    DDC 599.75/9--dc23/eng/20240129
LC record available at https://lccn.loc.gov/2023059416
LC ebook record available at https://lccn.loc.gov/2023059417

CONTENTS

SOUTH AFRICAN CHEETAHS

of Hwange National Park

Hwange National Park is the largest national park in Zimbabwe, a country in the southern part of the continent of Africa. At 5,863 square miles (15,185 square kilometers), the park is bigger than the state of Connecticut.

Flat grasslands and forests blanket Hwange. Hundreds of animal **species** make their homes here, including elephants, Cape buffalo, giraffes, antelope, zebras, lions, and cheetahs. Tens of thousands of visitors come each year to see the wildlife.

It's a warm day in mid-February. A female cheetah lounges in the shade of an acacia tree. In the distance, elephants roam and a steenbok races across the grass. A troop of baboons ambles by.

The cheetah pads the ground and shifts uncomfortably. Something momentous is about to happen. For 93 days, her cubs have been developing inside her. Her muscles contract, and she rises to her feet. It is time to give birth.

Mantle

Shortly after birth, a thick, soft, grayish mane appears on cubs' backs. This wispy coat may help the defenseless cubs blend into the grass. Some experts think it also helps regulate the cats' body temperature. The mantle starts to disappear after a few months.

CHAPTER ONE

LIFE BEGINS

Cheetahs are large, wild cats. Adults are around six feet long (almost two meters) and can weigh more than 140 pounds (64 kilograms). These lanky animals have long, slender legs and tawny coats covered in black spots. Each animal's pattern is unique, so no two cheetahs are identical. Bold black stripes run from the inner corners of their eyes down to the sides of the mouth.

Most cheetahs live in the grasslands of southern or eastern Africa. There, the tall, dry grasses offer camouflage for their golden coats.

Cheetahs are carnivores. Their favorite foods are gazelles and antelope, but they also eat warthogs and smaller creatures such as hares. Cheetahs stalk their prey as they move slowly, silently, and invisibly through the grass. They draw closer and closer. When they are close enough, cheetahs unleash their speed and sprint fast enough to overtake their prey.

SOUTH AFRICAN CHEETAH MILESTONES

DAY ①

- Born
- Blind and toothless
- Dark coat with spots that blend in
- Weight: 8 ounces (227 grams)

Cheetahs are the fastest land animal on Earth. They can go from standing still to running 60 miles (96 km) per hour in just three seconds. Their tail acts like a rudder, helping them zigzag back and forth as they chase their prey. Usually, the race is over in less than a minute. By that time, they will have caught their prey or given up the chase. Running at top speed is hard on a cheetah's body. The animals can keep up that pace for only a very short time.

As apex predators, cheetahs rule their food chain. No other animals hunt adult cheetahs, although sometimes lions eat baby cheetahs. Apex predators help keep animal populations in balance. Without cheetahs, the number of gazelles and antelope would explode. The hoofed grazers would eat more grass, which would lead to more soil **erosion**. Cheetahs help keep the food chain and their entire ecosystem healthy.

Voice box

A cheetah can't roar, but it can purr. It is the only big cat with a voice box, or larynx, made entirely of bone. It lacks the stretchy ligament needed to produce a roar.

— FEATURED FAMILY —

Welcome to the World

In Hwange National Park, the female cheetah turns in circles. Soon, the first cub is born. It's inside an amniotic sac, a fluid-filled membrane. The mother licks her cub. This clears away the membrane and helps the cub breathe air. The cub moves and stretches and rolls. Twenty minutes later, a second cub is born. A third comes eleven minutes later. Less than two hours after labor began, four cheetah cubs rest on the ground. They are blind, toothless, and helpless. The exhausted mother curls her body around them.

Although cheetahs grow up to be fast, fierce hunters, newborn cubs are helpless. They weigh less than one pound (0.5 kg) at birth and are blind and weak. They rely on their mother for milk. A mother cheetah goes off to hunt on most days. She must eat regularly to stay strong and produce enough milk for her hungry cubs. While she hunts, the cubs stay alone in a den. The mother moves the cubs every few days as she travels around her hunting grounds. Soon, the cubs grow much stronger.

① WEEK

- Opens eyes
- Crawls
- Grows a pale mantle along back

⑥ WEEKS

- Teeth have erupted
- Follows mother on hunting trips
- Begins eating meat

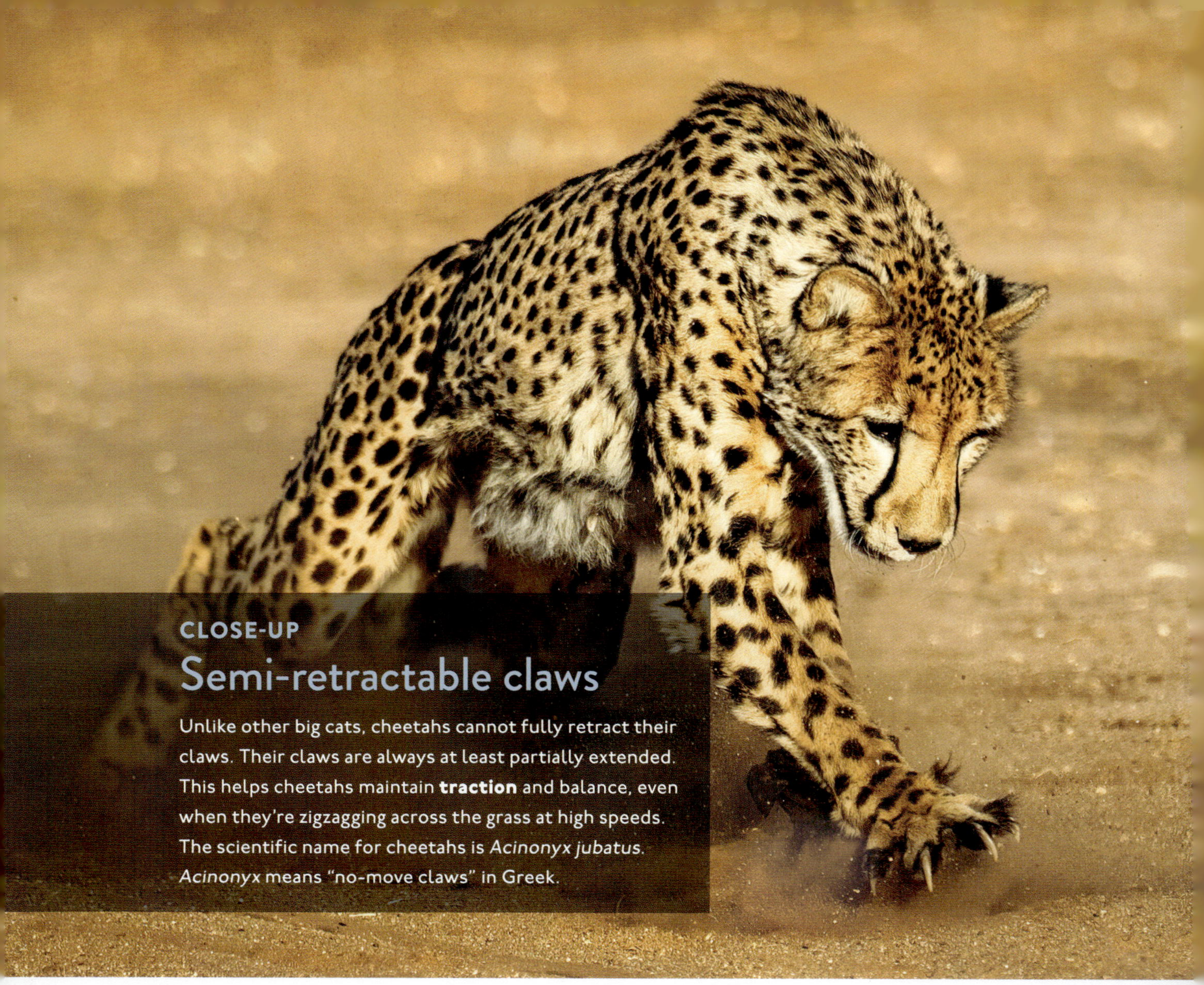

First Meal

The cubs cannot eat solid food. They have no teeth, and they depend solely on their mother's milk. At feeding time, the cubs curl up next to their mother. Each cub latches onto a nipple to drink nutrient-rich milk. This is the only food they will have for the first six weeks. After that, their teeth will erupt. Then they'll have the tools they need to start eating meat. Even then, they will continue to nurse until they are three to six months old.

The **cheetah** is the

FASTEST

land animal **on Earth.**

64 mph (103 kph): cheetah

43 mph (69 kph): greyhound

28 mph (45 kph): fastest human

(4) MONTHS

- ▸ Mantle starts falling out
- ▸ Nursing less
- ▸ Regularly leaves the den
- ▸ Mother and cubs are always together

CLOSE-UP
Eyesight
Cheetahs have strong vision. Their eyes
are set high on their heads. When they
gaze out over the savanna, they can
see a 210-degree view. By comparison,
humans can see only 140 degrees.

CHAPTER TWO

EARLY ADVENTURES

When they are about six weeks old, young cheetahs leave the den. They are eager to explore the savanna. But the cubs are still vulnerable, and the mother always stays nearby. There is danger everywhere. Lions, leopards, and hyenas all prey on cheetah cubs.

Predators aren't the only threat. Diseases are common. Anthrax, the deadliest disease for wild cheetahs, is caused by bacteria in the soil. A lack of food is also a threat. In some areas, nine out of every ten cheetahs die before adulthood. They most commonly die from illness or starvation. It's essential that cubs learn excellent hunting skills to increase their chance at survival.

By observing their mother, cubs learn to stalk, ambush, chase, and kill. They practice these behaviors by playing with their siblings. Young cubs are clumsy. But with each passing week, they grow faster, stronger, and more agile.

(6) MONTHS

- Watches mother closely to learn hunting behaviors
- Improves skills through rough play

(12) MONTHS

- Climbs trees to practice balancing
- Begins to participate in mother's hunts

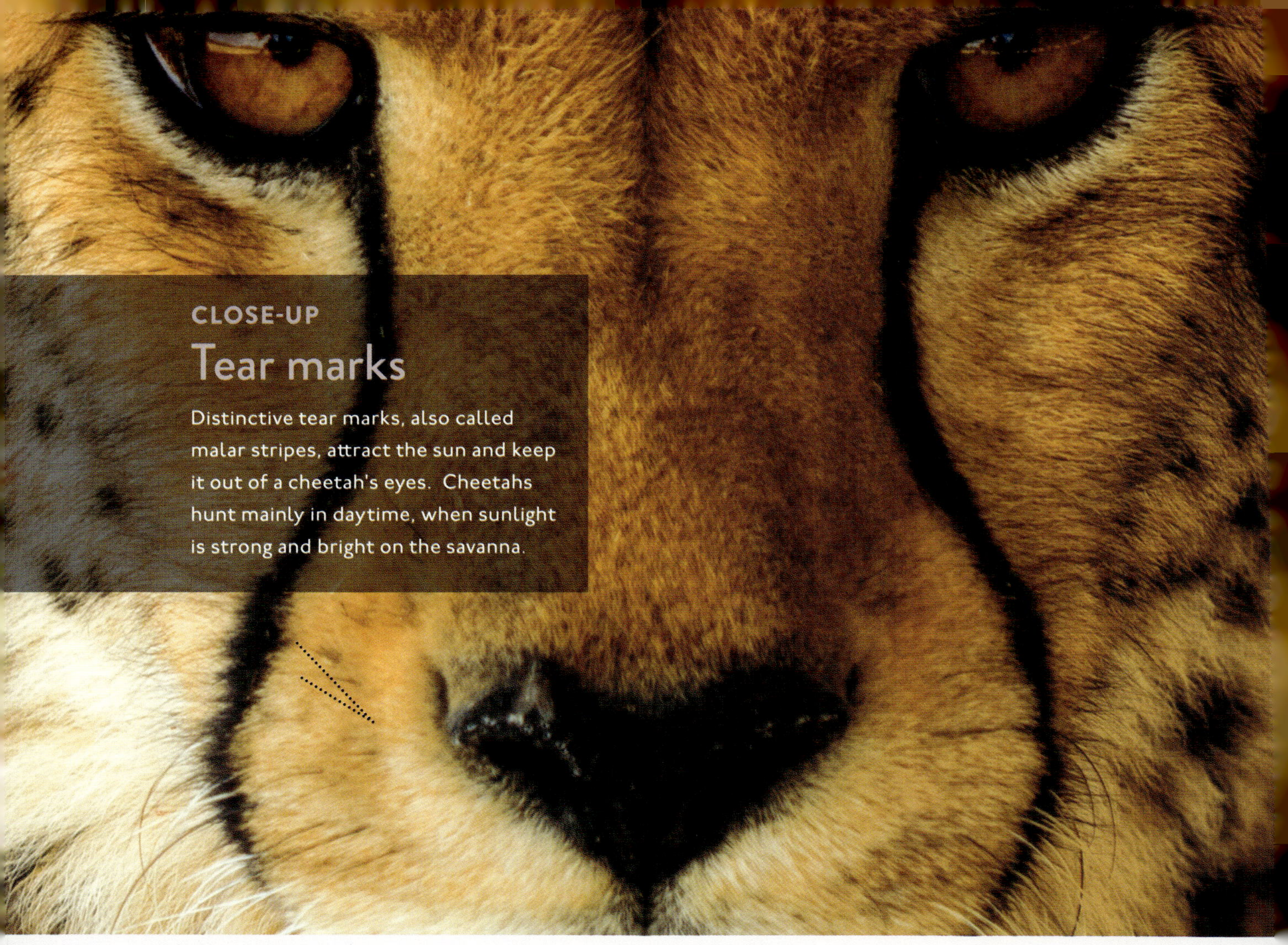

Look Who's Following Along

For six weeks, the mother left her cubs alone daily as she hunted. But now the cubs have developed enough to join her. Each day, they follow their mother to a new spot. She hunts while the cubs observe and explore. To-day, a hedgehog wanders close. One curious female cub sniffs it. It tucks its head and rolls into a ball of spiny quills. The cub growls and prods the hedgehog, but it doesn't run. The cub wants to chase something! She puts her paw on the hedgehog and sniffs deeply. Sharp quills stab her paw and nose. She learns that a hedgehog does not make a good playmate.

Cheetahs primarily eat gazelles and antelope. They rely on their exceptional eyesight and speed for hunting. The mother cheetah is the main hunter, but cubs jump in to help as much as they can.

As cubs grow older, their mother spends less time with them. When they are around one and a half years old, it is time for cubs to establish their own territories. The mother leaves to hunt one day and doesn't return. The siblings stick together for a few more months to perfect their hunting skills.

When the females are old enough to mate, the cubs separate. Male cubs often cooperate to hunt together and defend their shared territory. Each female lives alone, hunting across a broad home range.

(18) MONTHS

- Mother leaves permanently
- Stays close to siblings
- Hunts for own food with help of siblings
- Still perfecting hunting skills

Eye cones and rods

A cheetah's eyes are adapted for hunting in daylight. Compared to other big cats, their retinas contain more cone cells and fewer rod cells. This strengthens their daytime vision.

FEATURED FAMILY

Give It a Try

Mother often sits on a termite mound to spot prey. The female cub wanders away and finds an even higher vantage point—an acacia tree. Leopards and some other cats are excellent climbers. Cheetahs are not. Their non-retractable claws get dull from constantly pounding the ground. Dull claws don't grip well. Still, the cub scrambles up to a low branch. She watches elephants and springboks. Then she realizes that getting down is even more difficult than getting up. After a stumble and a leap, she is back on the ground.

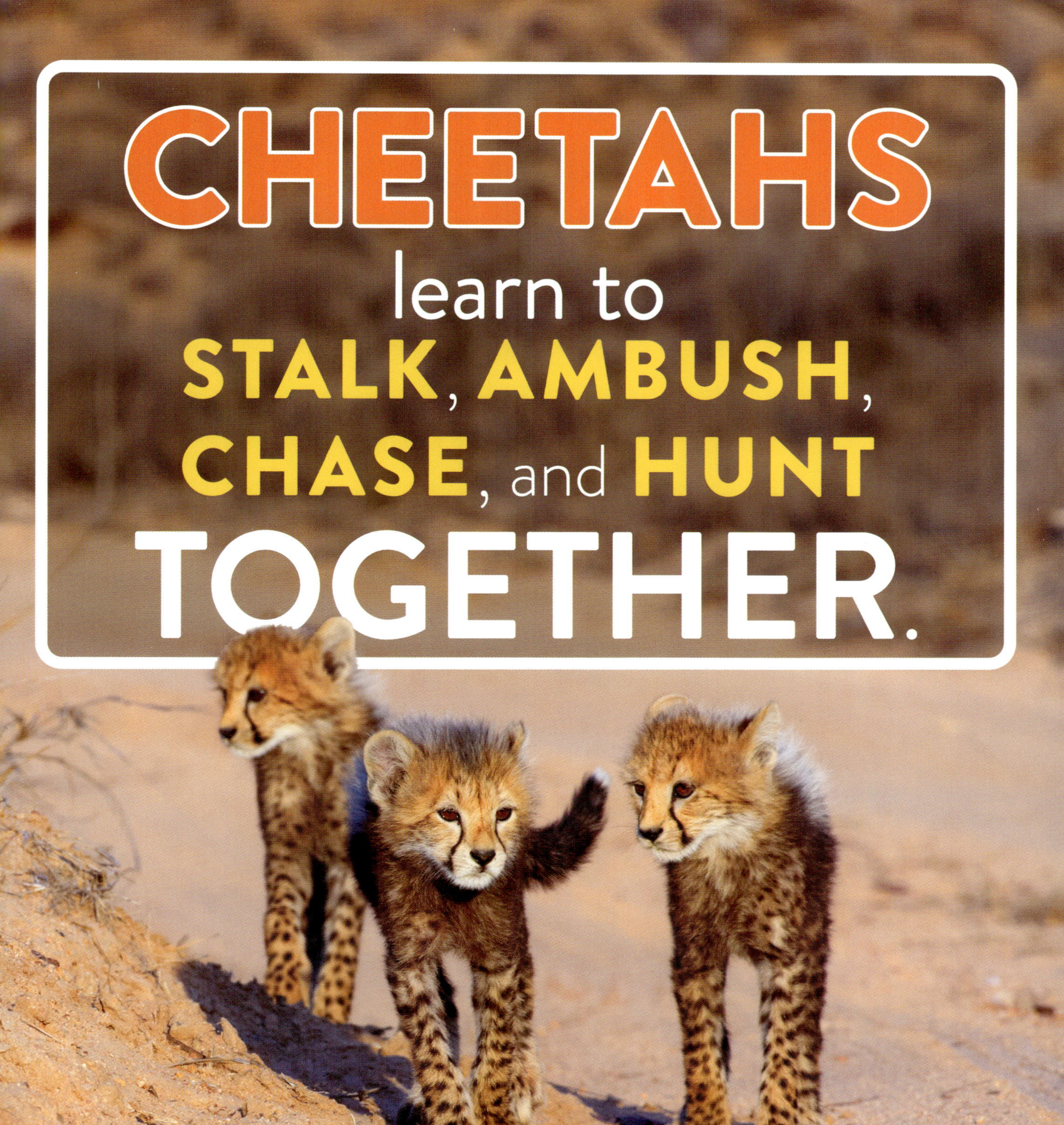

CHEETAHS
learn to
STALK, AMBUSH,
CHASE, and HUNT
TOGETHER.

(2) YEARS

▸ Splits up from siblings
▸ Females are ready to reproduce
▸ Females leave to live alone
▸ Male siblings may stay together

CLOSE-UP
Flexible spines
Of all the big cats, cheetahs have the longest spines compared to their overall length. Their spines are extremely flexible too. Their **vertebrae** bunch together and then expand with each stride, like a spring.

CHAPTER THREE

LIFE LESSONS

Most cheetahs live in sub-Saharan Africa. A tiny population of cheetahs remains in the Middle Eastern country of Iran. The big cats have adapted to many habitats, including grasslands, savannas, and deserts. One thing cheetahs need in any habitat is lots of open land. This allows them to use their most powerful hunting tool: their remarkable speed.

Cheetahs display many adaptations that make them perfectly built for their environments and their hunting techniques. They have long, slender bodies. Flexible spines let them make rapid maneuvers while chasing prey. Cheetahs even have extra-large nasal passages and lungs, so they get enough oxygen while sprinting. Their pale, spotted coats allow them to disappear in tall grasses. Compared to other big cats, cheetahs are light in weight. This helps them stalk quietly and sneak up on their prey. But a cheetah's teeth are small. That's why they don't want to fight with lions!

2.5 YEARS

- Males roam until they find a territory they can defend
- Females travel in ranges that overlap several male territories

3 YEARS

- Males live in territories up to 50 square miles (130 sq. km)
- Females live in home ranges of 300 square miles (800 sq. km) or more

At 20 to 24 months old, cheetahs reach sexual maturity. They are now old enough to breed. When a female is ready to mate, she urinates on trees and rocks. Males compete with each other to win the chance to mate with her. A cheetah's pregnancy lasts 90 to 95 days. Then she gives birth, usually to a litter of three to five cubs. A cheetah mother stays with her cubs for at least 12 to 18 months. That's how long it takes for cubs to be ready to live independently. Once her cubs are on their own, the mother cheetah breeds again and starts the cycle all over.

Adult female cheetahs without cubs live a solitary life. But adult males sometimes live in small **coalitions** with their male littermates. The brothers defend their territory and sometimes hunt together. In the wild, cheetahs live for 8 to 12 years.

This Is How It's Done

For six months, the cubs have watched their mother hunt. They've practiced stalking, chasing, and pouncing on each other throughout the summer. Now, at eight months old, they're ready for the next step. Mother chases down an impala, but she releases it before it dies. The cubs sprint after the weakened animal. They zig. They zag. The impala stumbles and falls. All four cubs clamber onto it and bite its neck to suffocate it. Their mother drags the impala to a hidden spot so that lions won't steal the cubs' first kill.

Nasal passages

Large nasal cavities mean less bone in a cheetah's skull. That makes its head lighter, so that it can hold its head and neck motionless as it sprints. The oversized nasal holes also allow the cheetah to take in plenty of oxygen. The extra oxygen gets used while the cat has its mouth closed, suffocating its prey.

3.5 YEARS
- Females give birth to first litter
- Most active in morning and evening

8 YEARS
- Females give birth to fourth litter
- Broken teeth make hunting harder

Cheetahs are important to their environment. They help prevent the overpopulation of grazing animals. Cheetahs mainly target weaker and slower animals. This strengthens the overall population of the prey animals. Keeping the number of grazing animals in check also helps protect the land. Overgrazing causes a weak root system in grasses and erodes the soil. In every habitat, living things work together for survival. When the population of one species grows too big or too small, all other species are affected too.

With their breathtaking speed and distinctive spots, cheetahs are majestic, beautiful animals. They're **ambassadors** for the African continent and a vital part of their habitats.

Fast-twitch muscles

An animal's muscle fibers can be fast-twitch or slow-twitch. Fast-twitch fibers can contract more quickly. A cheetah's leg muscles contain a high percentage of fast-twitch fibers. Cheetahs coil and release their leg muscles in a super-speedy way.

— FEATURED FAMILY —

Practice Makes Perfect

The young female kills a springhare. The small reddish-brown rodent weighs only five pounds (2 kg). After glancing around and seeing no large predators nearby, the hungry cub tears into her meal. She eats quickly, before lions or hyenas can come to snatch her prey. Since cheetahs are built for speed, not for fighting, they avoid other large predators. When the cub kills a young warthog the next day, she drags the heavier animal to a hiding place before eating.

MAJESTIC CHEETAHS
are ambassadors for the African continent.

12 YEARS
Females give birth to final litter   End of life

CHAPTER FOUR

HELPING CHEETAHS SURVIVE

Twelve thousand years ago, cheetahs roamed across Africa, Asia, Europe, and North America. Then the Ice Age hit. Most cheetahs didn't survive. Cheetah populations remained only in Africa and Asia. So few of the animals lived that the species lost much of its genetic diversity. That means cheetahs don't have a variety of **genes** to help them survive. It puts all cheetahs at greater risk for disease and death.

In the early 1900s, about 100,000 cheetahs traveled on the open grasslands. Today, just 7,000 cheetahs survive in the wild.

Low genetic diversity isn't the only problem. Cheetahs need large, open spaces for their territory. But humans have settled across 90 percent of the cheetahs' range. Open spaces have been replaced by ranches and farms.

Because of all these problems, cheetahs are incredibly fragile. Can the species be saved?

The Cheetah Conservation Fund (CCF) hopes so. This group in southern Africa focuses on saving the species by protecting both the animal and its habitat. In one program, it provides farmers with livestock guarding dogs (LGDs). Many farmers shoot and trap cheetahs because cheetahs occasionally kill livestock. LGDs protect farm animals by barking and scaring predators away. Farmers with LGDs are much less likely to kill cheetahs.

In another program, people harvest overgrown thornbushes. The plants can be turned into a kind of fuel and sold. This way, the land produces money for its owners. They don't have to farm or ranch the land to make money. Undeveloped land is a win for the owners and for the wildlife that depends on it.

Some zoos in the United States partner with organizations like CCF. These zoos work to educate the public, raise funds for cheetah conservation, and participate in breeding programs. The Cincinnati Zoo in Cincinnati, Ohio, raises money for CCF. In 2012, one of the zoo's cheetahs ran 100 meters in just 5.95 seconds. That made her the world's fastest land mammal on record.

Helping cheetahs survive takes effort from many groups and individuals. But all of them help make progress toward a single goal: to prevent the extinction of this beautiful animal.

SNAPSHOTS

The **South African cheetah** is the most numerous **subspecies**. About 4,300 live in southern Africa. That's more than half the world's remaining population of cheetahs.

Tourists from around the world come to Africa to watch wildlife like South African cheetahs. This **ecotourism** brings money to the region.

East African cheetahs are the biggest cheetahs. They live mainly in the Serengeti National Park and the Maasai Mara National Reserve.

The Maasai Mara National Reserve is famous for its annual Great Migration. Each year, millions of gazelles, zebras, and wildebeests travel across the plains.

Predators like East African cheetahs follow along behind the massive herds. They don't want grass. They're ready to dine on young, old, and sick animals.

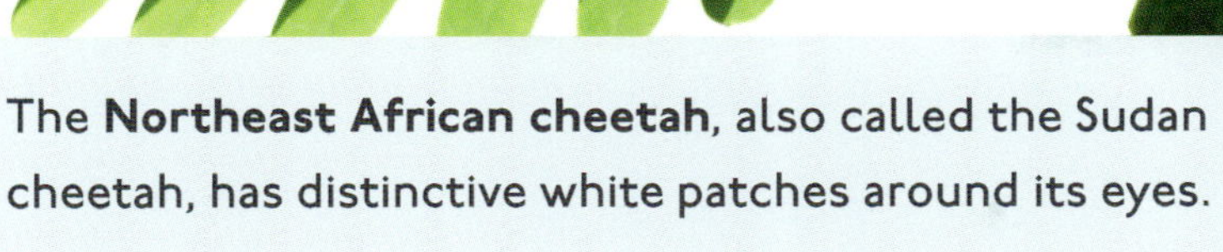

The **Northeast African cheetah**, also called the Sudan cheetah, has distinctive white patches around its eyes.

In cold climates, the Northeast African cheetah grows a fluffy winter coat. Fewer than 600 Northeast African cheetahs survive.

The **Northwest African cheetah** is elusive. Even scientists studying them hardly ever spot one.

The Northwest African cheetah lives in small numbers in the Sahara and Sahel regions. This cheetah has figured out how to survive in hot desert areas with no permanent water source.

The **Asiatic cheetah** is the rarest of all. It once roamed across central Asia. Fewer than 40 survive in the wild now.

The Asiatic cheetah lives in the deserts of Iran. It's smaller and lighter in color than the African cheetahs.

WORDS to Know

ambassadors	representatives of; symbols of
ambush	to attack by surprise from a hidden place
carnivores	animals that eat mostly meat
coalitions	a group that works together to achieve goals; an alliance
ecotourism	the practice of visiting a place to see its natural environments and wildlife
erosion	wearing away by wind and water
genes	building blocks that make up the DNA of living things and that determine their traits and characteristics
savanna	grassland with occasional trees; almost half of Africa is covered by savannas
species	a group of living things that have shared characteristics and that are able to reproduce with one another
subspecies	types or varieties of living things within a species
traction	the ability to stick to a surface while moving over it
vertebrae	bony segments that make up an animal's spine

LEARN MORE

Books

Johns, Chris, and Elizabeth Carney. *Face to Face with Cheetahs*. Washington, D.C.: National Geographic, 2008.

Montgomery, Sy. *Chasing Cheetahs: The Race to Save Africa's Fastest Cat (Scientists in the Field)*. New York: HarperCollins, 2017.

Winter, Steve. *The Ultimate Book of Big Cats: Your Guide to the Secret Lives of These Fierce, Fabulous Felines*. Washington, D.C.: National Geographic, 2022.

Websites

"About Cheetahs." Cheetah Conservation Fund. https://cheetah.org/learn/about-cheetahs

"Cheetah." National Geographic Kids. https://kids.nationalgeographic.com/animals/mammals/facts/cheetah

"Cheetah Cub Cam." Smithsonian's National Zoo & Conservation Biology Institute. https://nationalzoo.si.edu/webcams/cheetah-cub-cam

Documentaries

Cheetah: The Price of Speed. Earth Touch, 2011.

Miller, Alan. *Cheetahs Against All Odds*. PBS America, 2008.

Scholey, Keith, and Alix Tidmarsh. *African Cats*. Walt Disney Studios Home Entertainment, 2011.

Visit

CHEETAH CONSERVATION FUND

This organization welcomes visitors for tours, meeting orphaned cheetahs, and watching cheetahs practice their hunting skills.

D2440, Otjiwarongo, Namibia, Africa

CINCINNATI ZOO

Watch a cheetah chase a lure in the Cheetah Encounter show.

3400 Vine St.

Cincinnati, OH 45220

SAN DIEGO ZOO

Check out the cheetah exhibit and keep an eye out for ambassador cheetahs strolling the grounds with a wildlife care specialist.

2920 Zoo Drive

San Diego, CA 92101

SMITHSONIAN'S NATIONAL ZOO & CONSERVATION BIOLOGY INSTITUTE

Meet the cheetahs at the zoo's Cheetah Conservation Station.

3001 Connecticut Ave. NW

Washington, DC 20008

INDEX